Dimensions M
Textbook PKA

MW01492966

Authors and Reviewers

Tricia Salerno

Pearly Yuen

Jenny Kempe

Cassandra Turner

Allison Coates

Consultant

Dr. Richard Askey

Singapore Math Inc.

Published by Singapore Math Inc.

19535 SW 129th Avenue
Tualatin, OR 97062
www.singaporemath.com

Dimensions Math® Textbook Pre-Kindergarten A
ISBN 978-1-947226-00-5

First published 2017
Reprinted 2019, 2020

Printed in China

Acknowledgments

Editing by the Singapore Math Inc. team.
Design and illustration by Cameron Wray.

Preface

The Dimensions Math® Pre-Kindergarten to Grade 5 series is based on the pedagogy and methodology of math education in Singapore. The curriculum develops concepts in increasing levels of abstraction, emphasizing the three pedagogical stages: Concrete, Pictorial, and Abstract. Each topic is introduced, then thoughtfully developed through the use of exploration, play, and opportunities for mastery of skills.

Features and Lesson Components

Students work through the lessons with the help of five friends: Emma, Alex, Sofia, Dion, and Mei. The characters introduce themselves in Pre-K and continue to appear throughout the series. They give instructions, hints, and ideas.

Chapter Opener

Each chapter begins with an engaging scenario that stimulates student curiosity in new concepts. This scenario also provides teachers an opportunity to review skills.

Lesson

Engaging pictures draw the students into the concept of each lesson.

Exercise

A pencil icon ▬▬▬▬▶ at the end of the lesson links to additional practice problems in the workbook.

Review

A review of chapter material provides ongoing practice of concepts and skills.

Note: There are additional lesson components in the teacher's guide: Explore, Learn, Play, and Extend.

Emma   Alex   Sofia   Dion   Mei

Contents

Chapter	Lesson	Page

Chapter	Lesson	Page

Chapter	Lesson	Page
Chapter 7 **Numbers to 10** **— Part 2**	Chapter Opener	113

Chapter 1

Match, Sort, and Classify

Look and talk.

Hi, my name is Emma.
My favorite color is red.
Use your red crayon to match
the things that are the same.

Objective: Match objects that are the same and recognize the color red.

I'm Dion and my favorite color is blue.
Use your blue crayon to circle the blue things.

Objective: Recognize the color blue and circle blue objects.

Little Boy Blue, come blow your horn.
The sheep's in the meadow.
The cow's in the corn.
Where is that boy who looks after the sheep?
He's under the haystack fast asleep.

This is my favorite nursery rhyme.
See if you can figure out why.

Objective: Recognize the color blue.

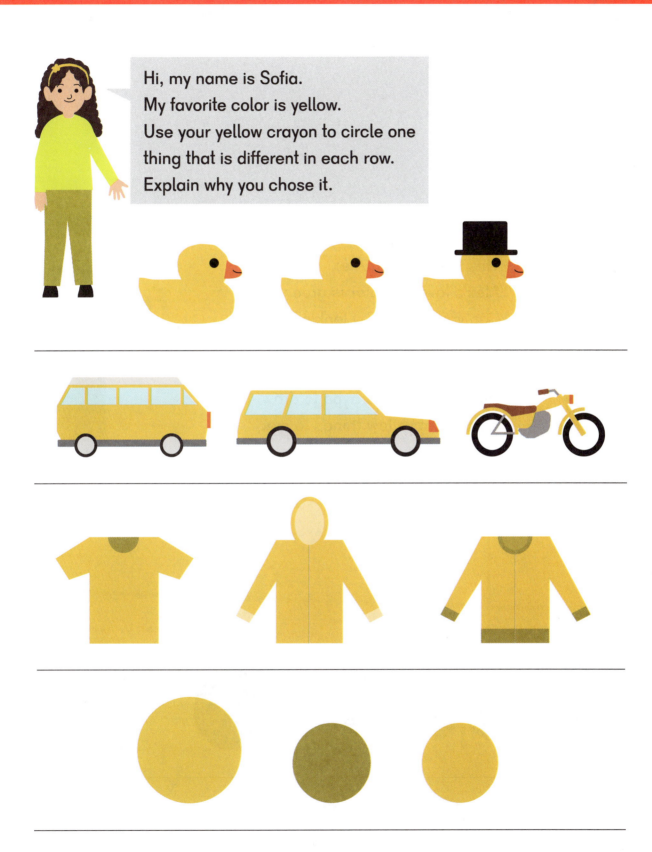

Hi, my name is Sofia.
My favorite color is yellow.
Use your yellow crayon to circle one thing that is different in each row.
Explain why you chose it.

Objective: Identify objects that are different.

This is my friend, Alex.
Alex's favorite color is green.
Can you tell that by looking at him?
Use your green crayon to draw
a line from green things to Alex.
Use your yellow crayon to draw
a line from yellow things to me.

Objective: Recognize the colors yellow and green.

Exercise 2 • page 3

My name is Mei.
Please help me sort my toys by
matching each toy to a basket.

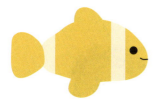

blue

red

yellow

Objective: Sort and classify objects by color and review the colors red, blue, and yellow.

Let's play a game!

The game is called "Red Light, Green Light."

Objective: Play a game to review the colors red and green.

Exercise 3 • page 5

1-3 Color Review

Some things are hard and some are soft. My teddy bear is soft. My t-ball bat is hard.

Objective: Explore textures and learn the words "soft" and "hard."

Circle all of the hard objects.

Objective: Classify the texture of objects as "soft" or "hard."

Objective: Explore textures and learn the word "rough."

This piece of sandpaper is rough enough to make the bumpy wood smooth.

This piece of satin is so smooth.

Objective: Classify objects by texture and learn the words "bumpy" and "smooth."

Exercise 5 • page 9

1-5 Rough, Bumpy, and Smooth

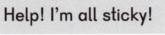

Objective: Explore textures and learn the words "sticky" and "grainy."

Circle the sticky things with your red crayon and circle the grainy things with your green crayon.

Sugar

GLUE

S

Objective: Classify objects by texture.

Exercise 6 • page 11

1-6 Sticky and Grainy

Objective: Use a familiar story to recognize the color orange and learn the words "big," "little," "small," and "large."

Papa Bear is big.

Baby Bear is little.

Papa Bear's chair was too large for Goldilocks.

Baby Bear's small chair was just right.

Match each bear to his chair.

Then match each spoon to a bowl.

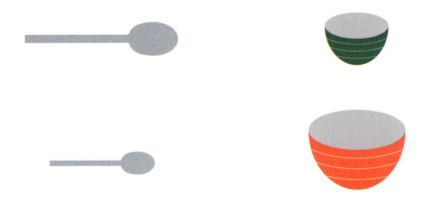

Objective: Match objects by size.

Exercise 7 • page 13

1-7 Size — Part 1

Objective: Compare objects by size.

Draw a big one.

Objective: Use drawing to review size.

Objective: Discuss reasons for sorting objects into two groups.

forks

spoons

Draw lines to sort these things into their box.

Objective: Sort objects into two groups and justify the sort.

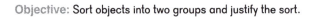

Exercise 9 • page 17

1-9 Sort Into Two Groups

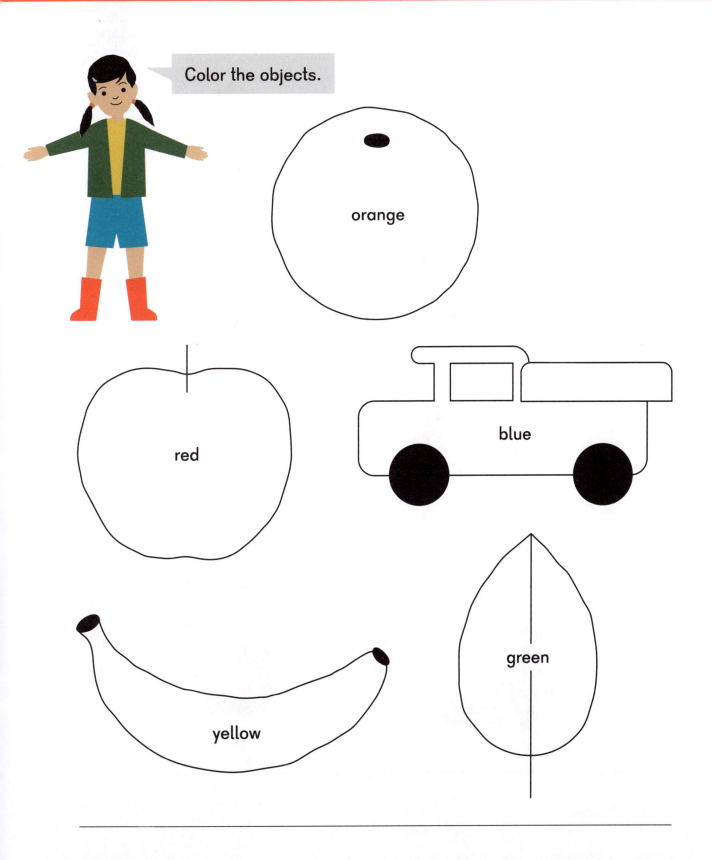

Color the objects.

orange

red

blue

yellow

green

Objective: Practice.

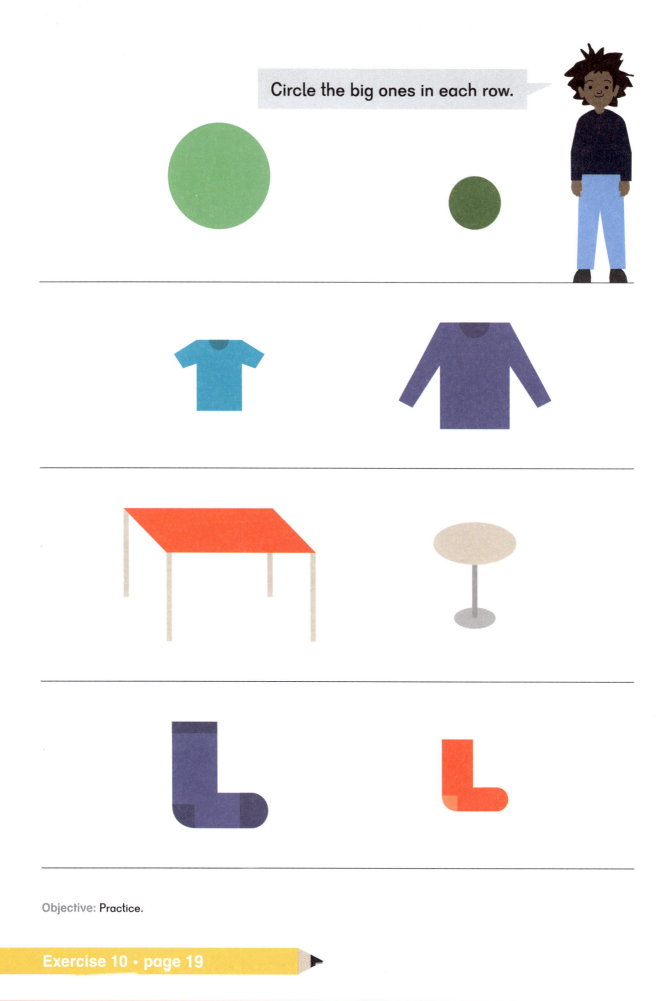

Circle the big ones in each row.

Chapter 2

Compare Objects

Objective: Compare objects by size and learn the words "bigger than," "biggest," "smaller than," and "smallest."

These are some of my toys.
Color the smallest one in each row green.
Color the biggest one in each row yellow.

Objective: Compare objects by size.

Exercise 1 • page 21

Objective: Compare objects by length and learn the words "long," "longer than," "longest," "short," "shorter than," and "shortest."

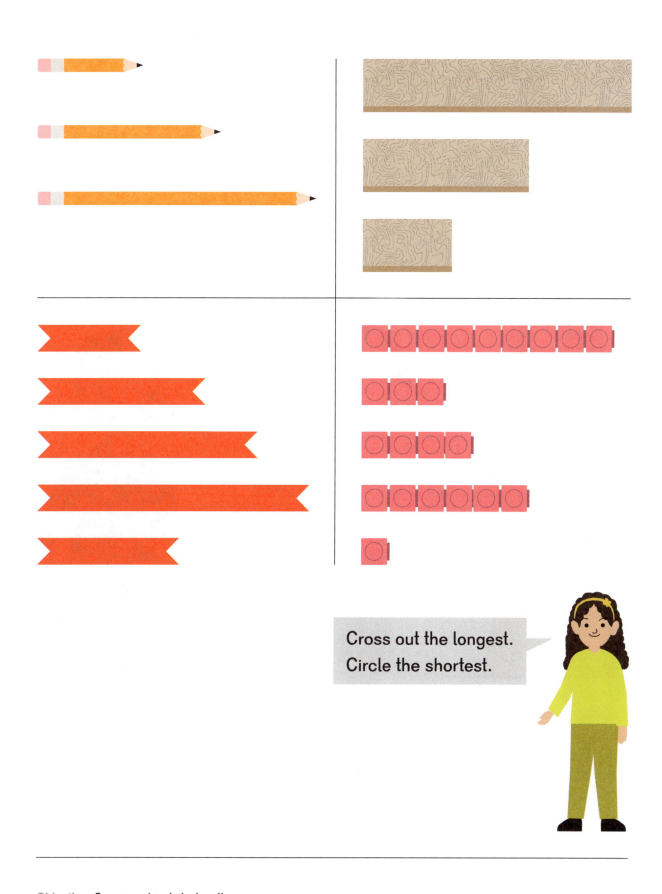

Cross out the longest.
Circle the shortest.

Objective: Compare objects by length.

Exercise 2 · page 23

I am taller than Alex.
I am shorter than Mei.
Mei is the tallest.

Objective: Compare objects by height and learn the words "tall," "taller than," and "tallest."

Color the tallest one yellow.
Color the shortest one orange.

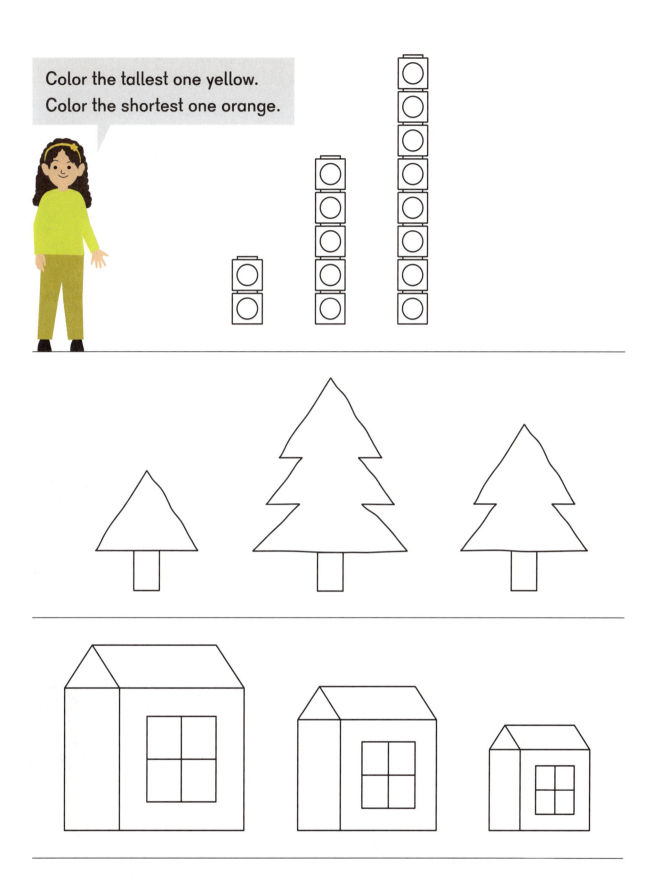

Objective: Compare objects by height.

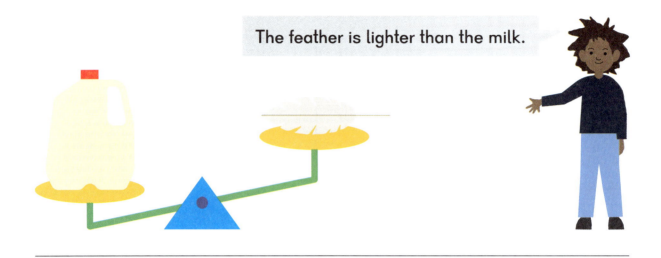

The feather is lighter than the milk.

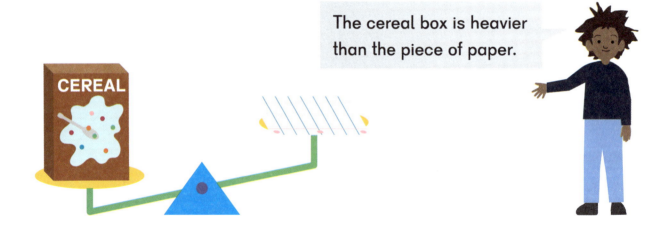

The cereal box is heavier than the piece of paper.

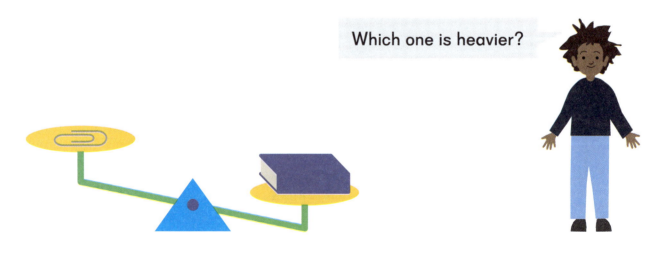

Which one is heavier?

Objective: Compare objects by weight and learn the words "heavy," "heavier than," "heaviest," "light," "lighter than," and "lightest."

I love to help with grocery shopping. Circle the heaviest one in each row.

Objective: Compare objects by weight.

Exercise 4 · page 27

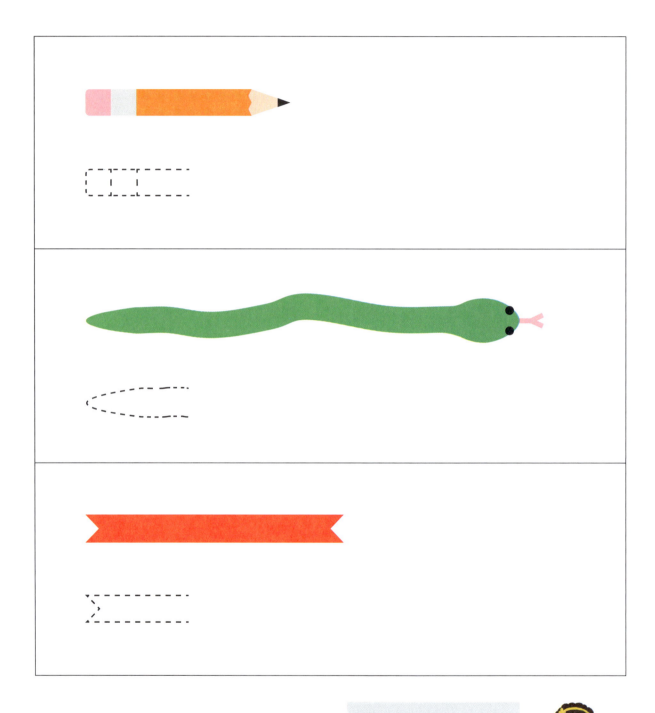

Draw a longer one.

Objective: Practice.

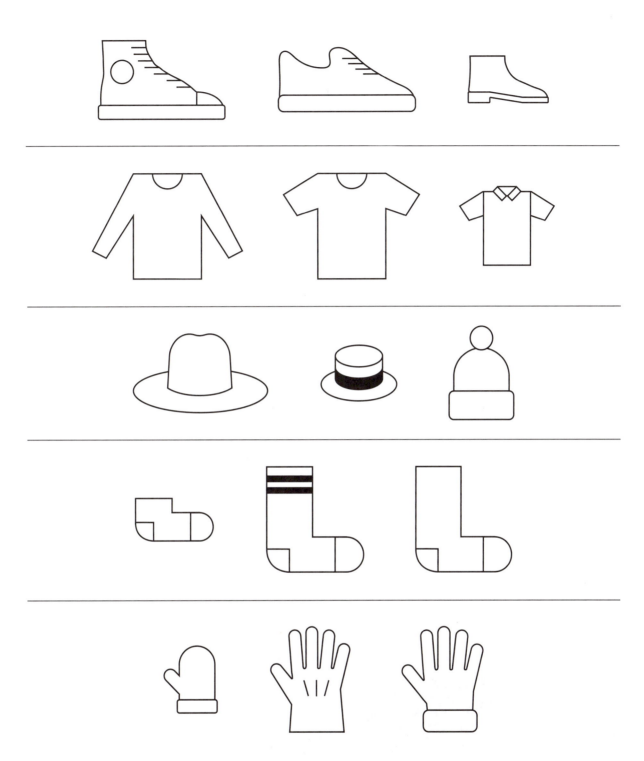

Color the smallest one.

Circle the lighter thing on each balance scale.

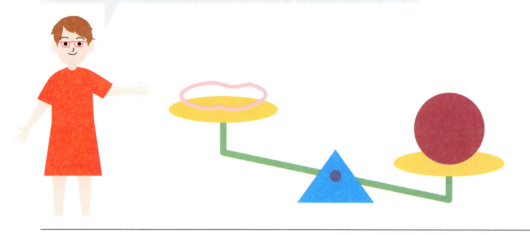

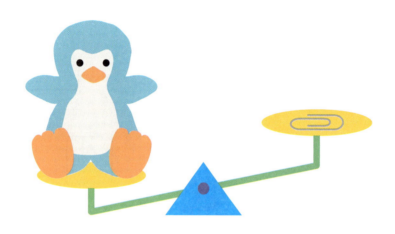

Objective: Practice.

Exercise 5 • page 29

2-5 Practice

Chapter 3

Patterns

Sofia and I go to gymnastics. Let's do a warm-up!

Objective: Create AB patterns using movement.

Mei and I take dance lessons.
Let's dance!

Objective: Create AB patterns using movement.

Exercise 1 • page 31

Let's clap!

Objective: Create AB, AAB, and ABB patterns using sound.

Objective: Create AB, AAB, and ABB patterns using sound.

Exercise 2 • page 33

I love making patterns with my train cars.
Use linking cubes to make the same color patterns I made.
The first pattern is black and white.

Color the linking cubes below to show the patterns that you made with your linking cubes.

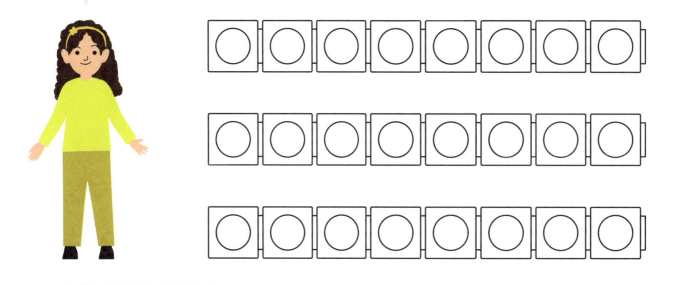

Objective: Create patterns using linking cubes and recognize the colors black and white.

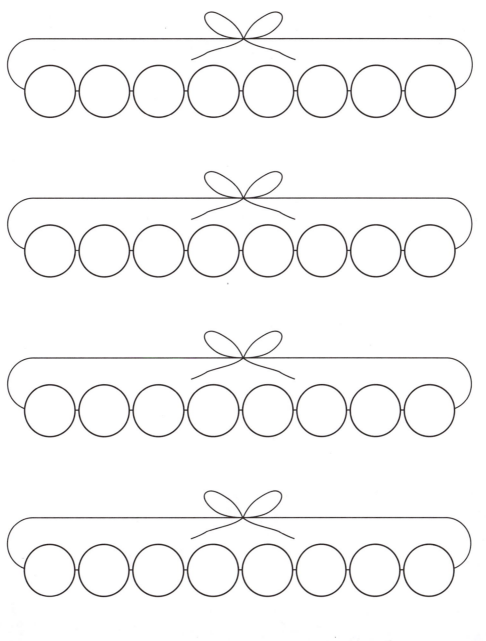

Color the beads to show some of your patterns.

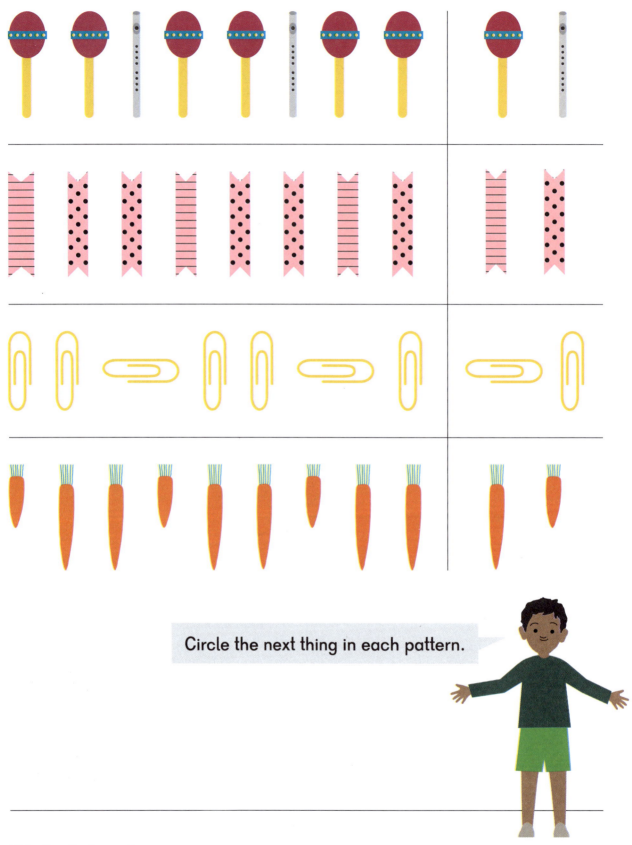

Circle the next thing in each pattern.

Objective: Continue patterns.

Exercise 3 • page 35

3-3 Create Patterns

What comes next?

Objective: Practice.

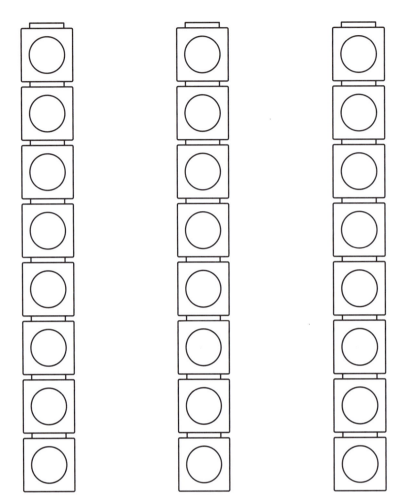

 Color each tower using 2 different colors to create a pattern.

Objective: Practice.

Exercise 4 • page 37

3-4 Practice

Chapter 4

Numbers to 5 — Part 1

Five fat peas in a pea pod pressed
(children hold hand in a fist)
One grew, two grew, so did all the rest.
(put thumb and fingers up one by one)
They grew and grew
(raise hand in the air very slowly)
And did not stop
Until one day the pod went POP!
(children clap hands together)

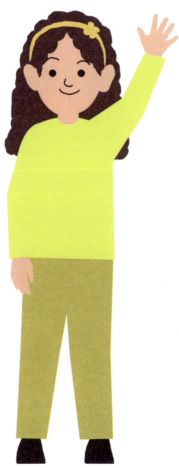

Objective: Count 1 to 5 by rote.

I like to count
My own fingers.
I have 5 on each hand.
1 – 2 – 3 – 4 – 5
1 – 2 – 3 – 4 – 5
1 – 2 – 3 – 4 – 5
5 fingers on my hand.
I like to count
My own cute toes.
I have 5 on each foot.
1 – 2 – 3 – 4 – 5
1 – 2 – 3 – 4 – 5
1 – 2 – 3 – 4 – 5
5 cute toes on my foot.

Objective: Count 1 to 5 by rote.

1 – 2 – 3 – 4 – 5 – I like to dance.
Let's shake and jive.

1 – 2 – 3 – 4 – 5 – I like to swim.
Let's learn to dive.

Objective: Count 1 to 5 by rote.

Lesson 3
Count Back

Five little monkeys jumping on the bed
One fell off and bumped his head.
Mama called the doctor and the doctor said,
"No more monkeys jumping on the bed!"

Four little monkeys jumping on the bed
One fell off and bumped his head.
Mama called the doctor and the doctor said,
"No more monkeys jumping on the bed!"

Three little monkeys jumping on the bed
One fell off and bumped his head.
Mama called the doctor and the doctor said,
"No more monkeys jumping on the bed!"

Two little monkeys jumping on the bed
One fell off and bumped his head.
Mama called the doctor and the doctor said,
"No more monkeys jumping on the bed!"

One little monkey jumping on the bed
One fell off and bumped his head.
Mama called the doctor and the doctor said,
"No more monkeys jumping on the bed!"

Objective: Count back from 5 to 1 by rote.

Five little ducks
Went out one day
Over the hills and far away.
Mama Duck said,
"Quack, quack, quack, quack."
Only four little ducks came back.

Four little ducks
Went out one day
Over the hills and far away.
Mama Duck said,
"Quack, quack, quack, quack."
Only three little ducks came back.

Three little ducks
Went out one day
Over the hills and far away.
Mama Duck said,
"Quack, quack, quack, quack."
Only two little ducks came back.

Two little ducks
Went out one day
Over the hills and far away.
Mama Duck said,
"Quack, quack, quack, quack."
Only one little duck came back.

One little duck
Went out one day
Over the hills and far away.
Mama Duck said,
"Quack, quack, quack, quack."
No little ducks came waddling back.

No little ducks
Went out one day
Over the hills and far away.
Sad Mama Duck said,
"Quack, quack, quack, quack."
All five ducks came waddling back.

Sing along if you know the words.

Objective: Count back from 5 to 1 by rote.

Objective: **Review counting from 1 to 5 and from 5 to 1.**

Objective: Review counting from 1 to 5 and from 5 to 1.

Objective: Count 1 object.

Color the 1 circle orange.

Objective: Count 1 object.

Exercise 1 • page 39

Objective: Count 2 objects with one-to-one correspondence.

Draw 2 big circles and color each of them 2 different colors.

Objective: Count 2 objects with one-to-one correspondence.

Exercise 2 • page 41

Objective: Count up to 3 objects with one-to-one correspondence.

Exercise 3 • page 43

Color the tower that has two squares in it purple.
Color the tower that has three squares in it yellow.
Color the tower that has four squares in it green.

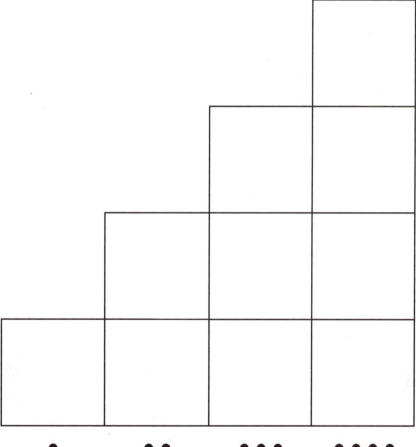

Objective: Count up to 4 objects with one-to-one correspondence and recognize the color purple.

Draw a line from the five-frame card to the set with the same number of cats.

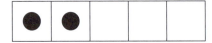   

Objective: Match five-frame representations of numbers 1 to 5 with a set of objects.

Match each set of t-ball bats to the five-frame card showing the same number.

Objective: Match five-frame representations of numbers 1 to 5 with a set of objects.

Exercise 5 • page 49

Let's play musical notes as we
count the fingers on our left hands.
How many fingers do we have on our left hand?

Objective: Count up to 5 objects with cardinality and recognize the color pink.

Old MacDonald had a farm,
1 – 2 – 3 – 4 – 5.
And on this farm he had 5 chickens,
1 – 2 – 3 – 4 – 5.
With a cluck, cluck, here and a cluck, cluck there.
Here a cluck, there a cluck, everywhere a cluck, cluck.
Old MacDonald had a farm,
1 – 2 – 3 – 4 – 5.

Old MacDonald had a farm,
1 – 2 – 3 – 4 – 5.
And on this farm he had 5 cows,
1 – 2 – 3 – 4 – 5.
With a moo, moo, here and a moo, moo there.
Here a moo, there a moo, everywhere a moo, moo.
Old MacDonald had a farm,
1 – 2 – 3 – 4 – 5.

Old MacDonald had a farm,
1 – 2 – 3 – 4 – 5.
And on this farm he had 5 ducks,
1 – 2 – 3 – 4 – 5.
With a quack, quack, here and a quack, quack there.
Here a quack, there a quack, everywhere a quack, quack.
Old MacDonald had a farm,
1 – 2 – 3 – 4 – 5.

Objective: Count up to 5 objects with cardinality.

Playing store is fun. Count the fruits and vegetables.

Objective: Count up to 5 objects with cardinality.

Exercise 7 • page 53

Objective: Understand that a rearrangement of objects does not change the number of objects in the set, and recognize the color brown.

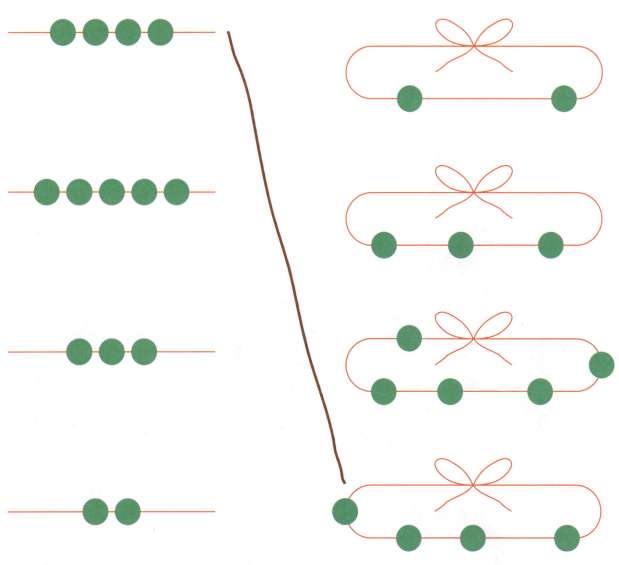

 Use your brown crayon to match the beads.
I did one for you.

Objective: Understand that a rearrangement of objects does not change the number of objects in the set, and recognize the color brown.

Exercise 8 • page 55

Objective: Understand that a rearrangement of objects does not change the number of objects in the set.

Use your black crayon to match the tower to the same number of blocks.

Objective: Understand that a rearrangement of objects does not change the number of objects in the set.

Exercise 9 • page 57

Color the 1 bead string pink.
Color the 3 bead string brown.
Color the 2 bead string black.
Color the 4 bead string green.

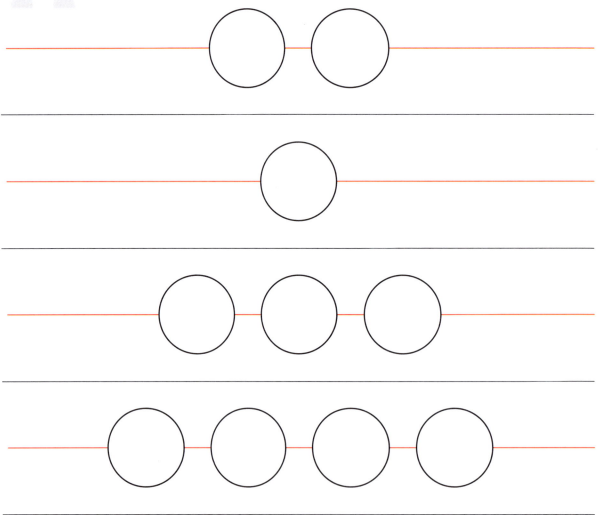

Objective: Practice.

Color in the number of boxes
to match the number of fruits.

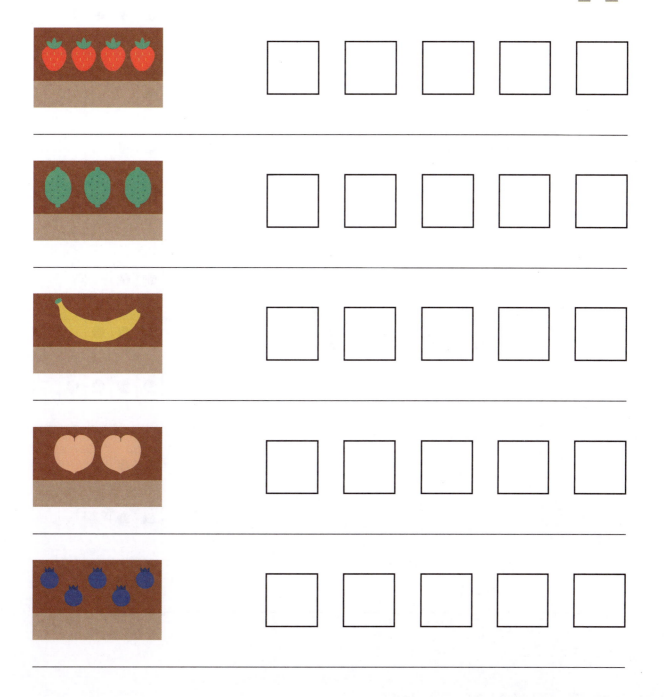

Objective: Practice.

Circle the five-frame card that goes with the number of train cars.

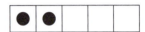

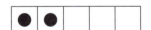

Objective: Practice.

Exercise 10 • page 59

4-14 Practice

Chapter 5

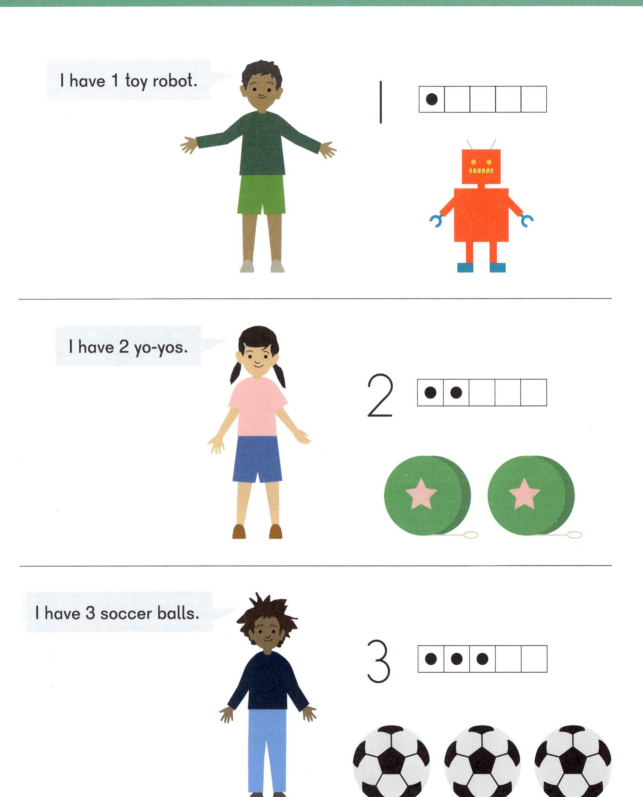

I have 1 toy robot.

I have 2 yo-yos.

I have 3 soccer balls.

Objective: Recognize numerals 1, 2, and 3.

2

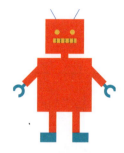

3

|

Draw a line to match.

Objective: Recognize numerals 1, 2, and 3.

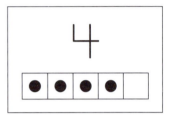

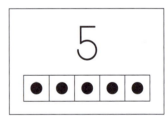

Draw lines to match the number of children to the number card.

Objective: Recognize numerals 1 to 5.

Let's use a number path to help us with our numbers.
I will hop to the number of toys I see.

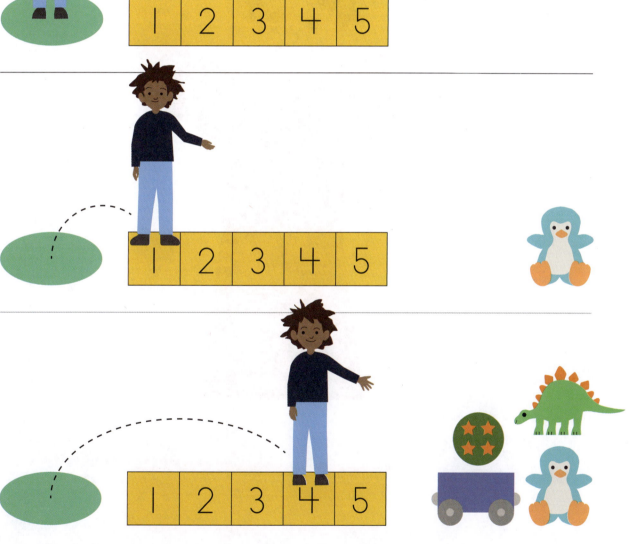

Objective: Recognize numerals 1 to 5 on a number path.

Let's use a five-frame card to help us with our numbers.
Let's count some things in the classroom.

Objective: Put counters on a five-frame card to match the number of objects in a set of up to 5 objects.

3

5

2

4

1

Draw a line to match the objects to the correct number.

Objective: Match a given numeral, 1 to 5, to a set containing that number of objects.

Exercise 4 · page 71

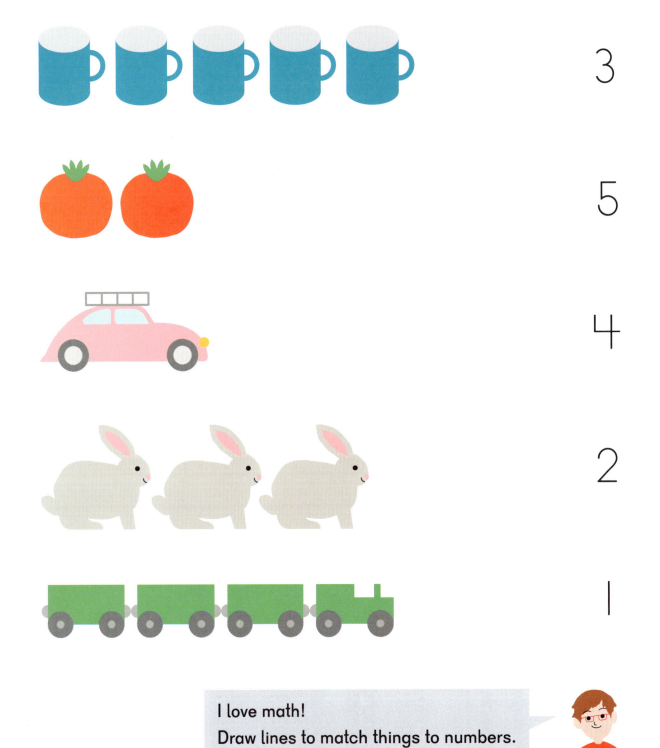

3

5

4

2

|

I love math!
Draw lines to match things to numbers.

Objective: Match a given numeral, 1 to 5, with a picture containing that number of objects.

Drawing is fun!
Let's draw the number of squares for each number.
The first one is done for you.

1	☐
2	
4	
5	
3	

Objective: Match a given numeral, 1 to 5, with a picture containing that number of objects.

Exercise 5 • page 73

Roll and tell.

Objective: Recognize a number of objects, up to 5, without counting.

Exercise 6 • page 75

Objective: Recognize that rearranging a number of objects does not change the number of objects in the set.

Circle the groups that show 3.

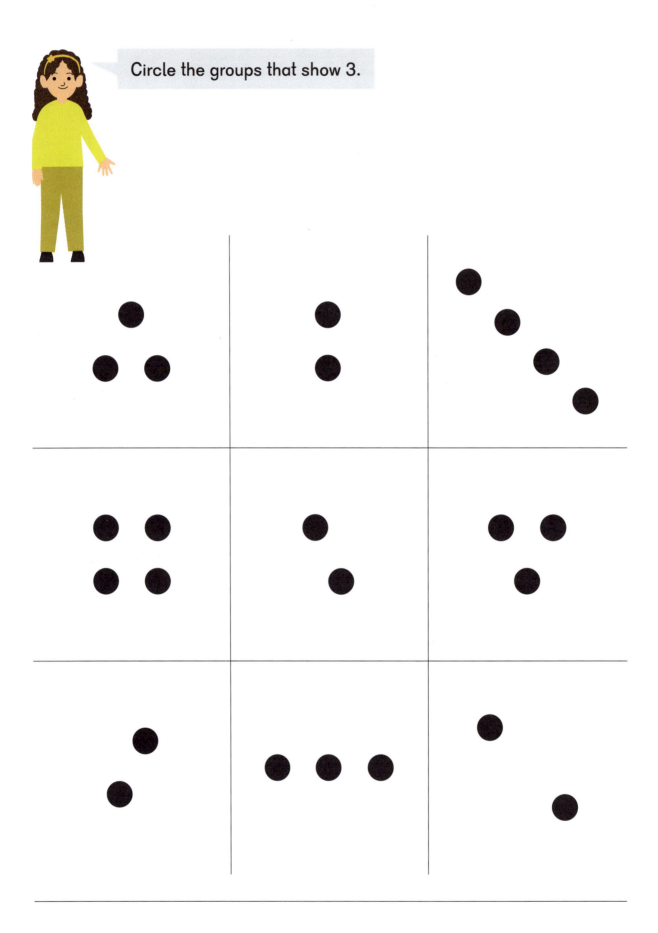

Objective: Recognize that rearranging a number of objects does not change the number of objects in the set.

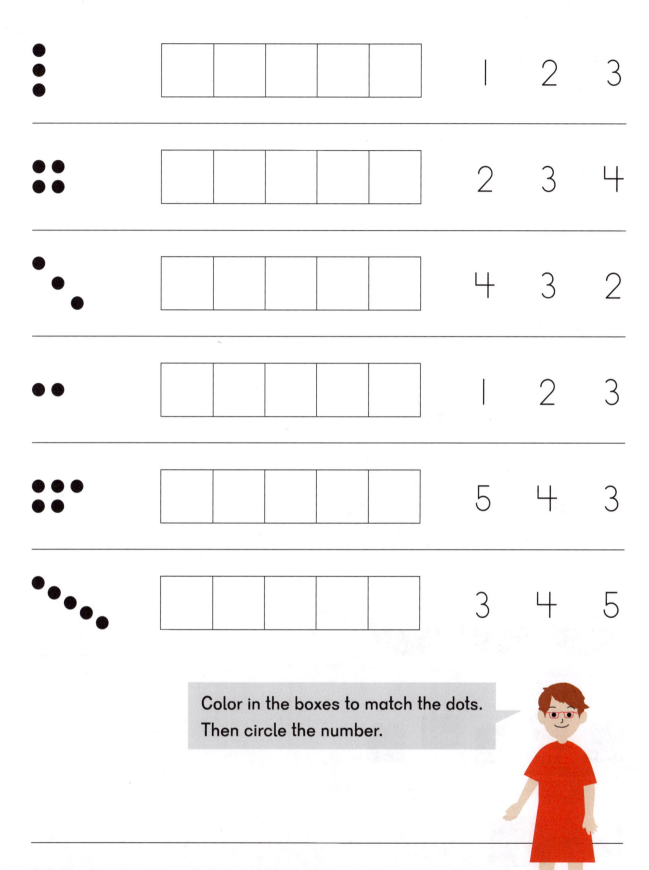

1   2   3

2   3   4

4   3   2

1   2   3

5   4   3

3   4   5

Color in the boxes to match the dots.
Then circle the number.

Objective: Recognize that rearranging a number of objects does not change the number of objects in the set.

Exercise 7 • page 77

Color in the boxes to show the number of things.

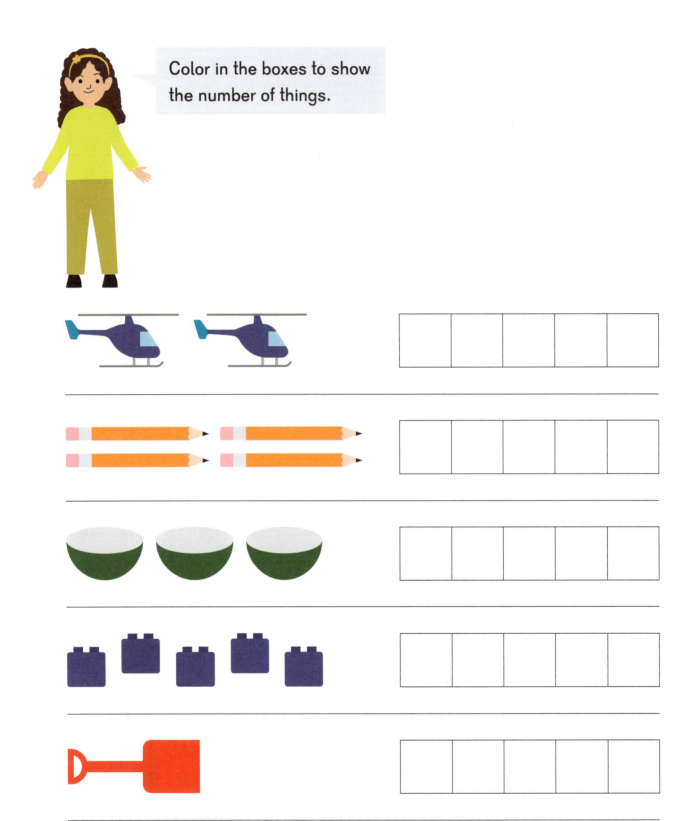

Objective: Practice.

 1 2 3 4 5

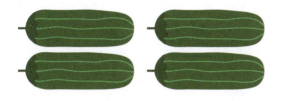 1 2 3 4 5

 1 2 3 4 5

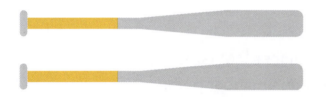 1 2 3 4 5

 1 2 3 4 5

Circle the number.

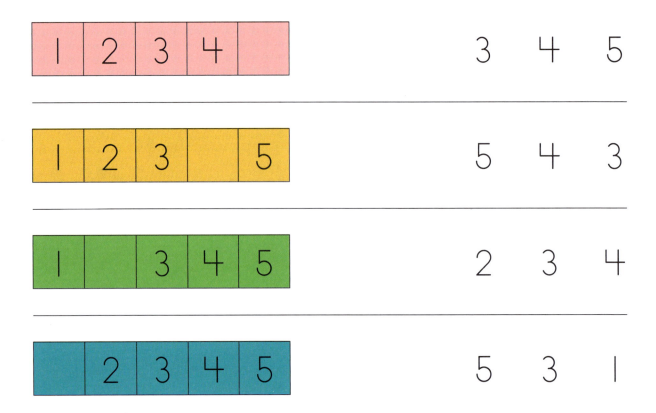

1 2 3 4 ☐	3 4 5
1 2 3 ☐ 5	5 4 3
1 ☐ 3 4 5	2 3 4
☐ 2 3 4 5	5 3 1

Circle the number that is missing in each number path.

Objective: Practice.

Exercise 8 • page 79

5-8 Practice

Chapter 6

Numbers to 10 — Part 1

89

Spot has 3 doggie biscuits.

Objective: Understand that a set with no objects in it is a set of zero.

Spot ate 1 doggie biscuit.
Now he has 2 doggie biscuits.

Spot ate another doggie biscuit.
Now he has 1 doggie biscuit.

Oh no! Spot ate all of his doggie biscuits.
He has none left.

There is a number for none.
We call that number zero.
Spot has zero doggie biscuits.

Objective: Understand that a set with no objects in it is a set of zero.

This is what zero looks like.

   0   4   2

   2   3   5

   4   3   0

   1   2   3

Circle the number of doggie biscuits in each bowl.

   4   5   2

Objective: Understand that a set with no objects in it is a set of zero.

Exercise 1 • page 81

One two three four fi-ive si-ix seven
Eight nine ten, I counted to ten.
One two three four fi-ive si-ix seven
Eight nine ten, I counted again.

Let's sing!

Objective: Count to 10 by rote.

Do you like to go fishing?
Let's learn a rhyme about it!

1 – 2 – 3 – 4 – 5
Once I caught a fish alive.
6 – 7 – 8 – 9 – 10
Then I let it go again.
Why did you let it go?
Because it bit my finger so.
Which finger did it bite?
The little finger on my right.

Objective: Count to 10 by rote.

Objective: Count from 10 to 1 by rote.

Before blasting off, astronauts count back from 10.

10 − 9 − 8 − 7 − 6 − 5 − 4 − 3 − 2 − 1 − BLAST OFF!

Objective: Count from 10 to 1 by rote.

Objective: Count 1 to 7 in correct sequence.

We are going to start using ten-frame cards instead of five-frame cards now because we are learning greater numbers. The number after 5 is 6.

Objective: Count up to 6 objects with one-to-one correspondence.

Circle the baskets that have 6 vegetables in them.

Objective: Count up to 6 objects with one-to-one correspondence.

Exercise 2 • page 83

Sunday	Monday	Tuesday	Wednesday	Thursday	Friday	Saturday

There are 7 days in a week.

Let's count them and sing a song.

Sunday,
Monday,
Tuesday,
Wednesday,
Thursday,
Friday,
Saturday,
Then we start again.

Objective: Count up to 7 objects with one-to-one correspondence.

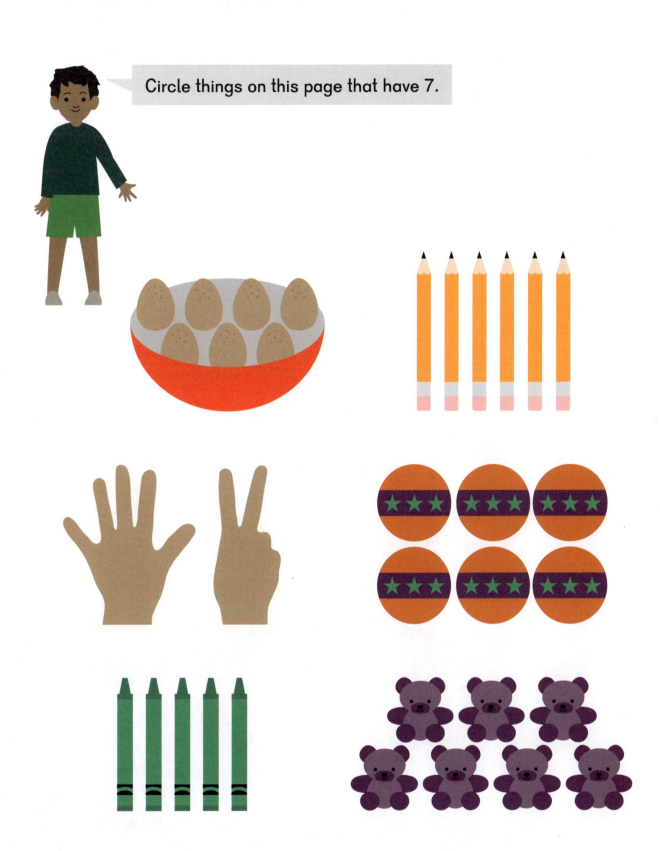

Circle things on this page that have 7.

Objective: Count up to 7 objects with one-to-one correspondence.

Exercise 3 • page 85

This octopus sure has a lot of arms!
Let's count them.

Objective: Count up to 8 objects with one-to-one correspondence.

Exercise 4 • page 87

There are a lot of cars on this train.
Let's count them!

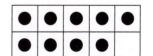

Objective: Count up to 9 objects with one-to-one correspondence.

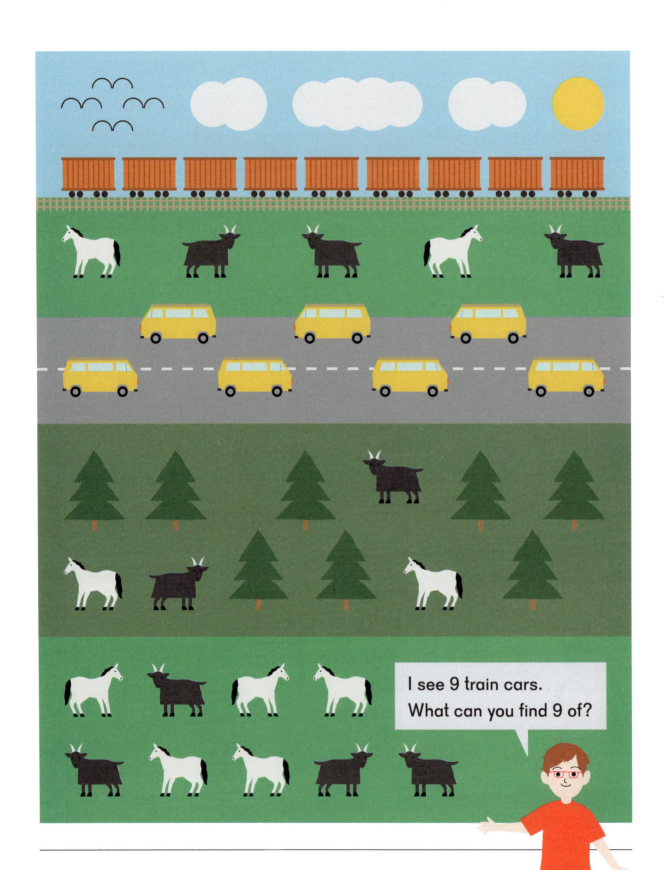

I see 9 train cars.
What can you find 9 of?

Objective: Count up to 9 objects with one-to-one correspondence.

Exercise 5 • page 89

6-9 Count Up to 9 Objects

Objective: Count up to 10 objects with one-to-one correspondence.

One, two, buckle my shoe.
Three, four, shut the door.
Five, six, pick up sticks.
Seven, eight, lay them straight.
Nine, ten, a big fat hen.

Objective: Count up to 10 objects with one-to-one correspondence.

Exercise 6 • page 91

6-10 Count Up to 10 Objects — Part 1

Lesson 11
Count Up to 10 Objects — Part 2

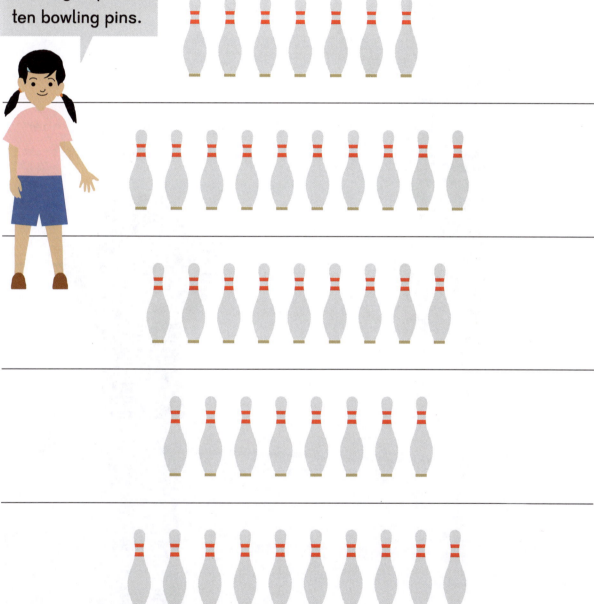

Bowling is so much fun! Circle groups of ten bowling pins.

Objective: Count up to 10 objects with one-to-one correspondence.

Exercise 7 • page 93

How many fingers am I showing?

I'm showing the same number of fingers a different way!

Objective: Count up to 10 objects with cardinality.

How many cans are in each group?
Color in the correct part of each ten-frame card.

Objective: Count up to 10 objects with cardinality.

Circle the nest with 0 birds.

Objective: Practice.

Circle the groups of 7.

Objective: Practice.

Color the same number of beads.

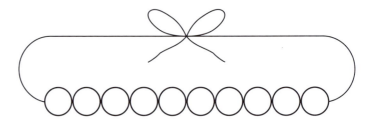

Objective: Practice.

Chapter 7

Numbers to 10 — Part 2

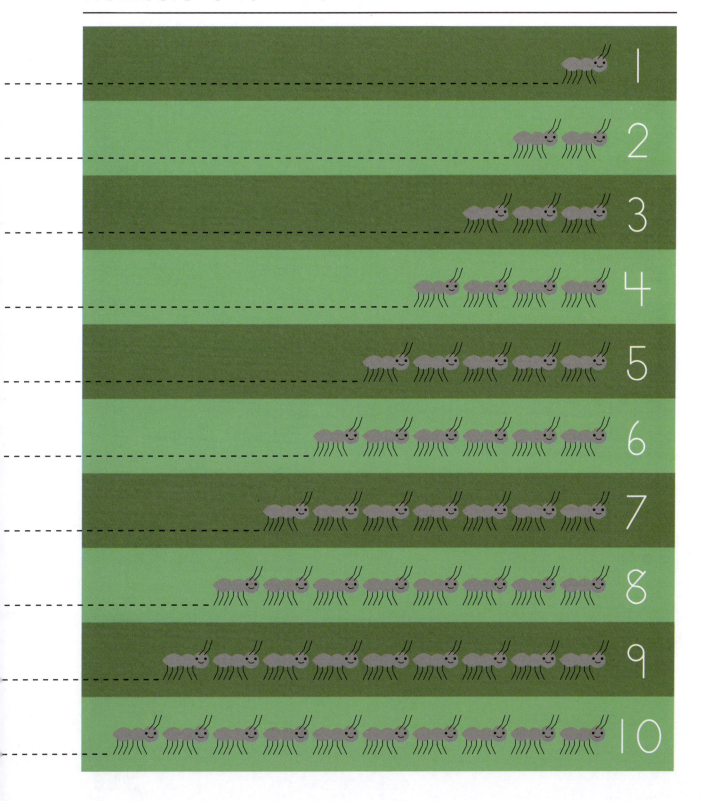

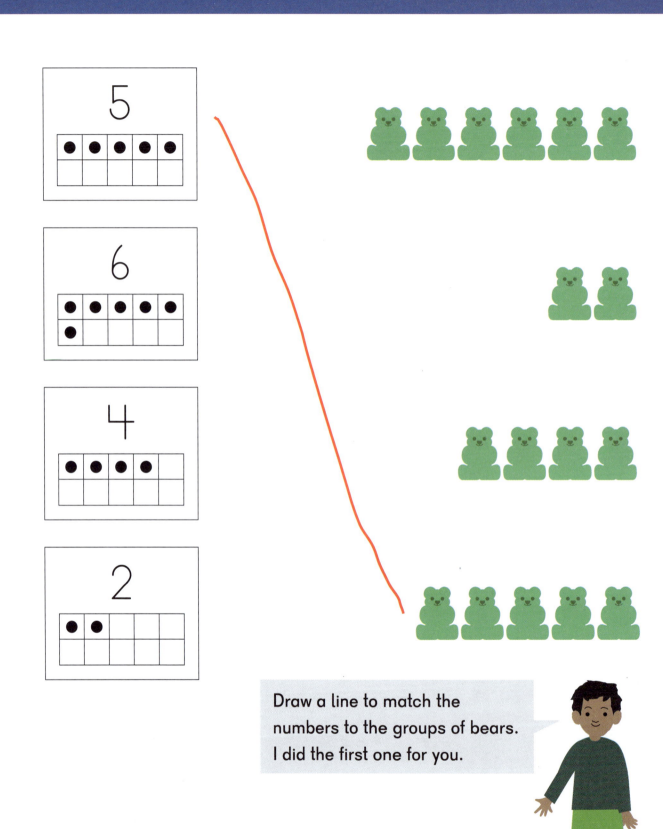

Draw a line to match the numbers to the groups of bears. I did the first one for you.

Objective: Recognize the numeral 6.

Circle the number of marbles in each group.

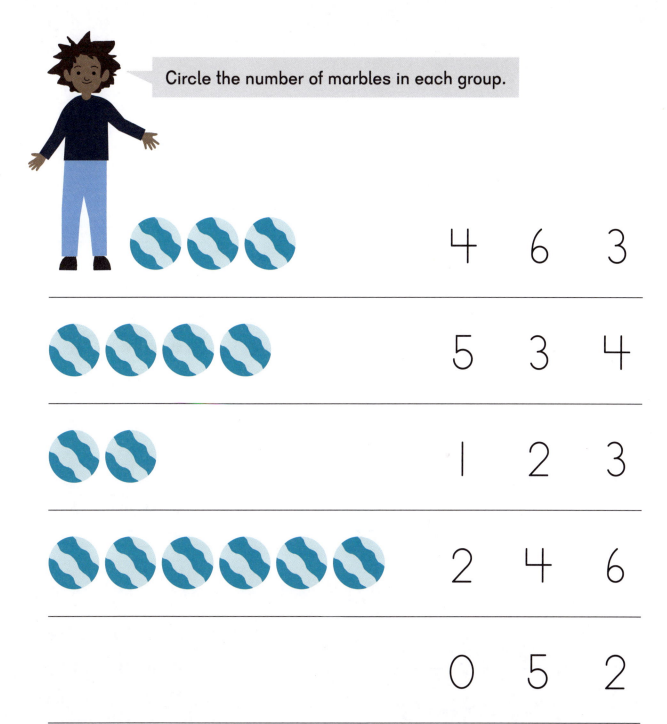

marbles			
●●●	4	6	3
●●●●	5	3	4
●●	1	2	3
●●●●●●	2	4	6
	0	5	2
●●●●●	6	5	1

Objective: Recognize the numeral 6.

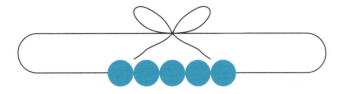
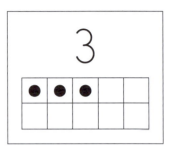

3

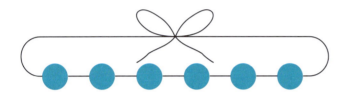
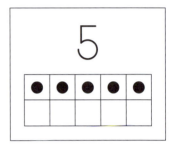

5

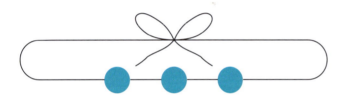
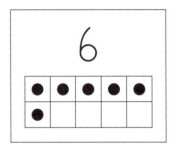

6

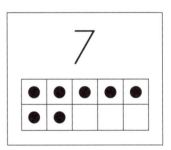

7

Match the necklace to the number.

Objective: Recognize the numeral 7.

 Circle all the 7s you see.

Objective: Recognize the numeral 7.

Exercise 2 • page 101

Match the peanuts to the numbers.
I did the first one for you.

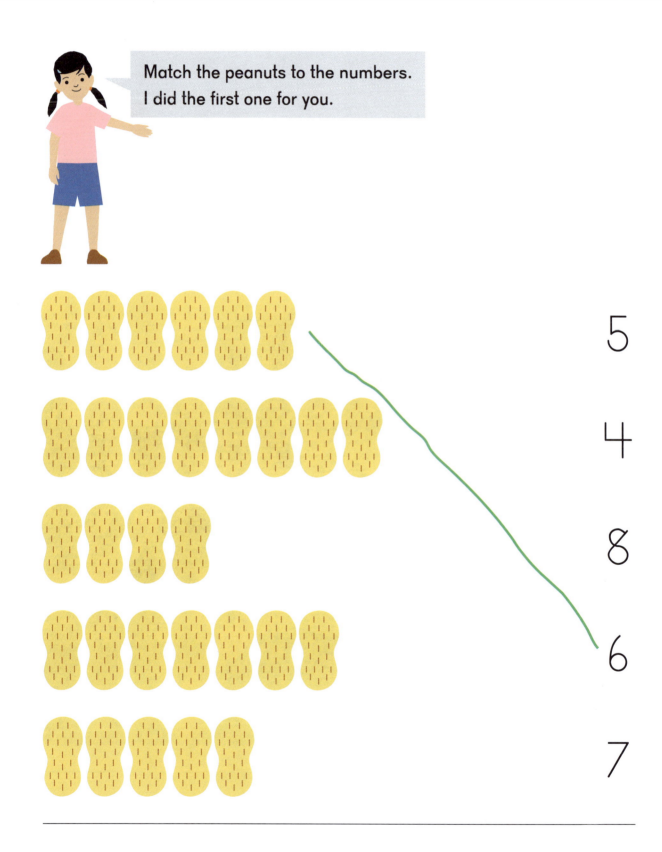

5

4

8

6

7

Objective: Recognize the numeral 8.

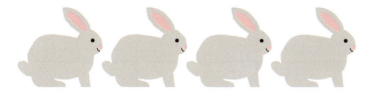 4   6   3

 8   7   6

 4   6   8

 7   6   5

 2   3   4

Circle the correct number.

Objective: Recognize the numeral 8.

| 4 | 6 | 8 | 5 | 3 |

Color the correct number of boxes.
I did the first one for you.

Exercise 3 • page 103

Objective: Recognize the numeral 8.

All of us are showing the number 9.
There are 9 of us here, including me.

Objective: Recognize the numeral 9.

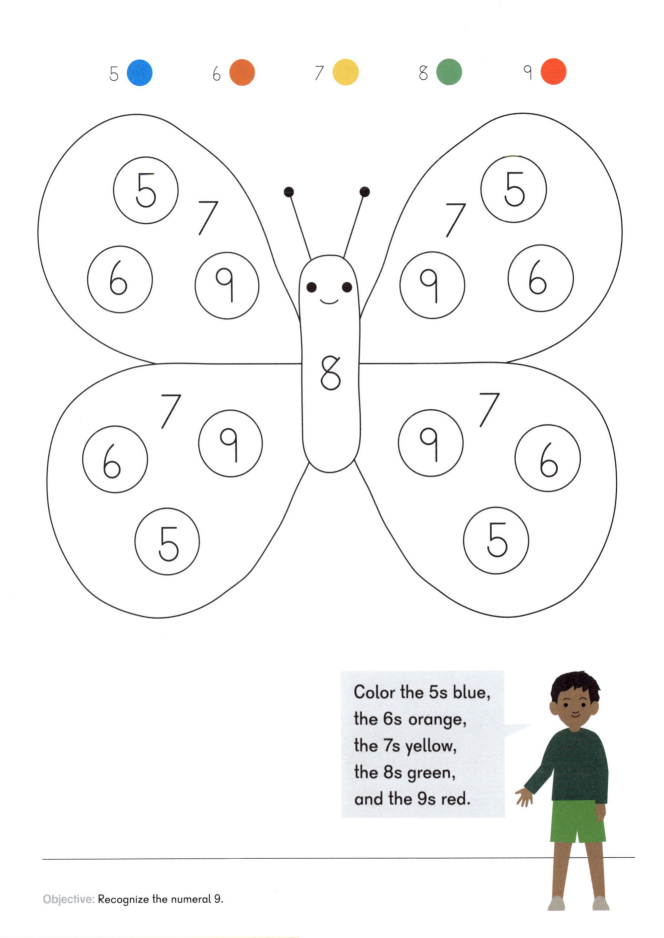

5 ● 6 ● 7 ● 8 ● 9 ●

Color the 5s blue,
the 6s orange,
the 7s yellow,
the 8s green,
and the 9s red.

Objective: Recognize the numeral 9.

Exercise 4 • page 105

1 cute kitten

2 and 3

I love kittens.

Don't you see?

4 cute kittens

5 and 6

I love kittens.

More than soccer kicks!

7 cute kittens

8 and 9

I love kittens.

Especially when they're mine!

10 cute kittens

They all are mine.

I love kittens.

They are mighty fine.

This is one of my favorite rhymes. Can you guess why?

Objective: Recognize the numeral 10.

 3   4   5

 1   8   10

 3   5   7

2   1   0

 9   8   7

 6   4   9

Circle the correct number.

Match the ten-frame card to the number.

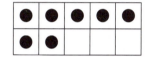

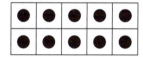

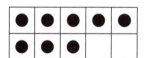

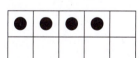

8

4

0

10

7

5

Objective: Recognize the numerals 0 to 10.

Match the crayons to the number.

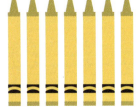

2

6

3

7

5

Objective: Recognize the numerals 0 to 10.

Exercise 6 • page 109

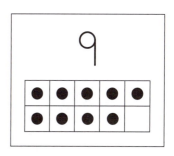

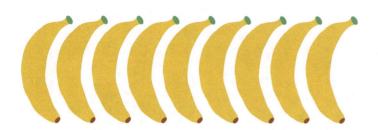

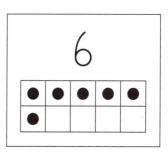

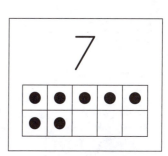

Draw a line from each group of bananas to the matching number card.

Objective: Match a given numeral 0 to 10 to a set containing that number of objects.

Exercise 7 • page 111

Lesson 8
Count and Match — Part 2

Objective: Match the number of objects in a picture to a numeral card.

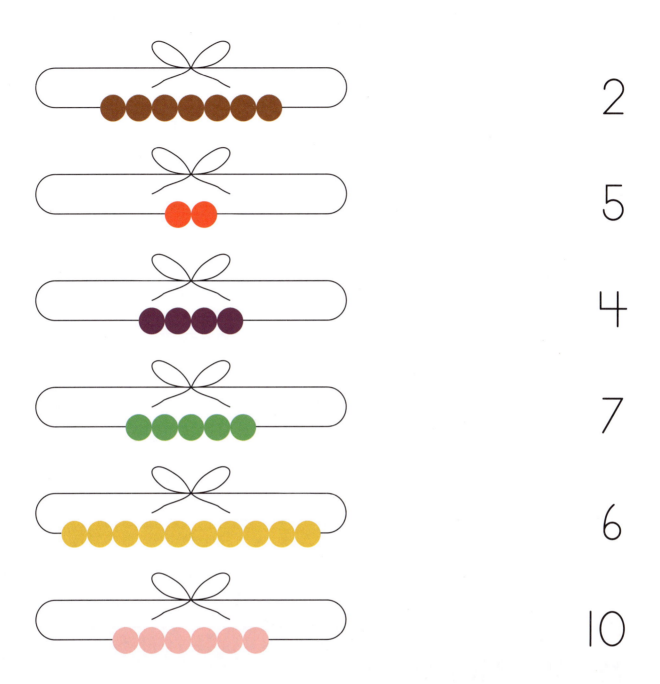

2

5

4

7

6

10

Match each necklace to the correct number.

Objective: Match a given numeral 0 to 10 with a picture containing that number of objects.

0

4

7

5

Draw the number of beads on the strings.

Objective: Draw the number of objects shown with a numeral.

Exercise 8 • page 115

7-8 Count and Match — Part 2

6

1

5

9

4

3

Match the flowers to the number.

Objective: Practice.

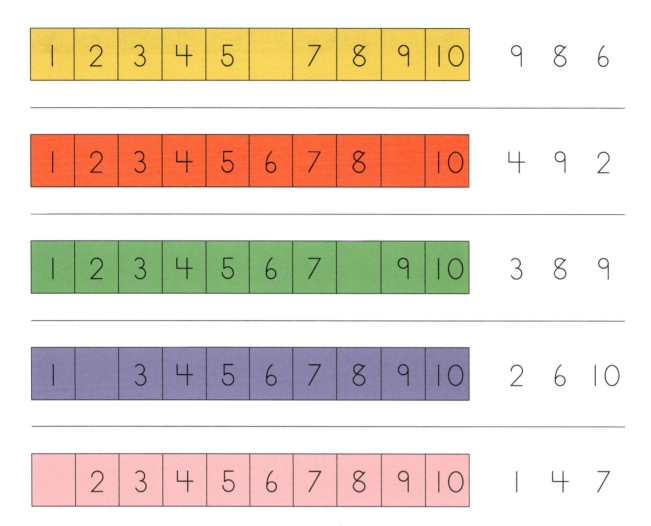

| 1 | 2 | 3 | 4 | 5 | | 7 | 8 | 9 | 10 | 9   8   6 |

| 1 | 2 | 3 | 4 | 5 | 6 | 7 | 8 | | 10 | 4   9   2 |

| 1 | 2 | 3 | 4 | 5 | 6 | 7 | | 9 | 10 | 3   8   9 |

| 1 | | 3 | 4 | 5 | 6 | 7 | 8 | 9 | 10 | 2   6   10 |

| | 2 | 3 | 4 | 5 | 6 | 7 | 8 | 9 | 10 | 1   4   7 |

Circle the number missing on each number path.

Objective: Practice.

Cross out enough balls so that there are 0 left.

Objective: Practice.

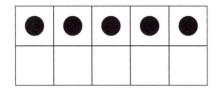   5   9   7

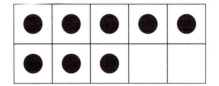   7   8   6

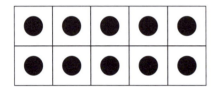   10   9   5

   5   8   9

Circle the number shown
on each ten-frame card.

Objective: Practice.

Blank

Blank